Alchemical Symbols

from

Medicinisch-chymisch-
und alchemistisches Oraculum, 1755.

Edited with explanations
by Adam McLean

Hermetic Studies Series No 10.
Kilbirnie

Hermetic Studies No. 10
25 Main Street
Kilbirnie KA25 7BX

Introduction

Adam McLean

From the beginning of alchemy, as can be seen though viewing the earliest manuscripts and books, alchemists have always used little graphic symbols, alike to modern day icons, to represent substances or aspects of their alchemical work. Of course many of these were idiosyncratic, even devised for a single piece of writing, but as alchemy matured through the centuries, a body of established alchemical symbols arose, many of which could be universally recognised across the whole of Europe. Many listings were printed in books and copied extensively in manuscripts.

Although alchemical literature went into sharp decline at the close of the 18th century, so that few books at all were issued in the 19th century, in the late 1700s the most comprehensive listing of alchemical symbols was published at Ulm in Germany in 1755. The title reads in English:

> A medical chemical and alchemical Oraculum in which is
> shown not only the symbols and abbreviation which occur
> in the recipes and books of the physicians and pharmacists
> as well as in the writings of the chemists and alchemists,
> but also a very rare manuscript of a certain realm.

This was reprinted in 1772, 1782 and 1783, so there must have been a continuing interest in owning a copy. The book consists of a series of thirty four full page engravings of 1829 symbols with the names in Latin and German overprinted in letterpress. There are 359 entries with multiple graphic images associated with each entry. Some ten years ago I placed these images onto my alchemy web site. A number of people then asked me to provide explanations for the obscure terms, but I was not immediately able to find the time to do this. At the end of 2013 I had the idea of issuing this in book form, and after a few months editorial work and research this has resulted in the present volume in my Hermetic Studies series.

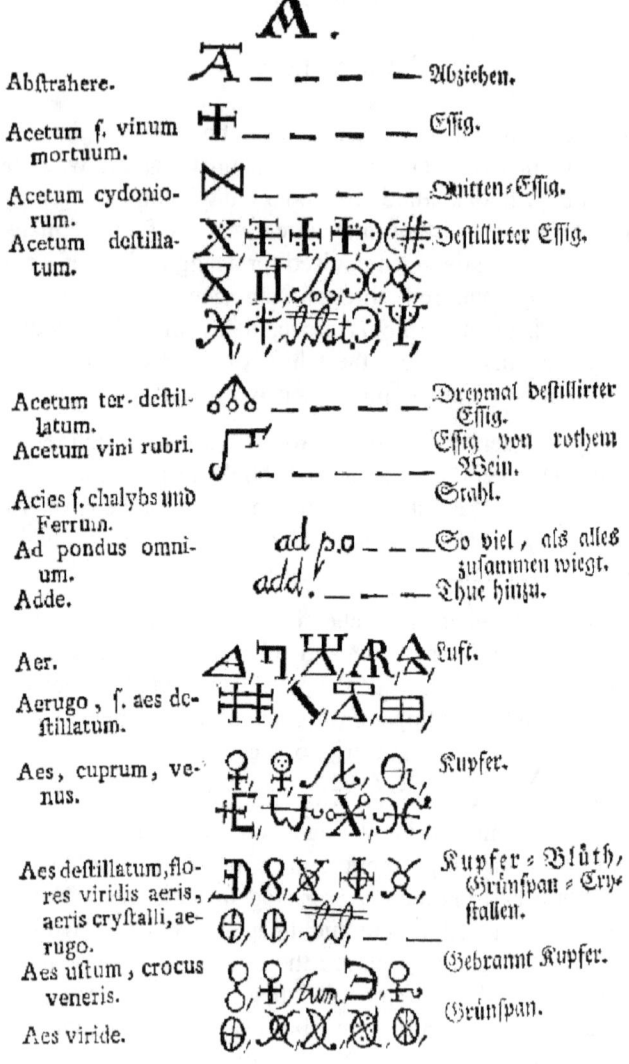

A.

Abſtrahere. — — — — — Abziehen.

Acetum ſ. vinum mortuum. — — — — — Eſſig.

Acetum cydonio-rum. — — — — — Quitten-Eſſig.

Acetum deſtilla-tum. — — — — — Deſtillirter Eſſig.

Acetum ter-deſtil-latum. — — — — Dreymal beſtillirter Eſſig.

Acetum vini rubri. — — — — — Eſſig von rothem Wein.

Acies ſ. chalybs und Ferrum. — Stahl.

Ad pondus omni-um. — So viel, als alles zuſammen wiegt.

Adde. — Thue hinzu.

Aer. — Luft.

Aerugo, ſ. aes de-ſtillatum.

Aes, cuprum, ve-nus. — Kupfer.

Aes deſtillatum, flo-res viridis aeris, aeris cryſtalli, ae-rugo. — Kupfer-Blüth, Grünſpan-Cry-ſtallen.

Aes uſtum, crocus veneris. — Gebrannt Kupfer.

Aes viride. — Grünſpan.

2

Abstrahere : To abstract or draw out. A process of extracting a substance from some ore or solution.

Acetum : See vinum mortuum. The vinegar made from sour wine.

Acetum cydoniorum : The vinegar made from quinces.

Acetum destillatum : Distilled vinegar, a strong acid.

Acetum ter-destillatum : Vinegar three times distilled. Very strong acetic acid.

Acetum vini rubri : Vinegar made from sour red wine.

Acies : See Chalybs und Ferrum. Steel. The sharp metal used for sword blades made from iron.

Ad pondus omnium : The whole weight, as much as it all together weighs.

Adde : Add.

Aer : Air.

Aerugo : See aes destillatum. Verdigris, the green rust, copper acetate $Cu(C_2H_5CO_2)_2$ that forms on copper when exposed to vinegar. It was commercially prepared from the 15th century as a pigment. It was often impure and contained the basic carbonate and the chloride of copper. It boils at a relatively low temperature so can be distilled and purified.

Aes, cuprum, Venus : Copper, the metal of Venus.

Aes destillatum, flores viridis aeris, aeris crystalli, aerugo : Verdigris.

Aes ustum. Crocus veneris : Burnt copper, probably a mixture of cuprous oxide Cu_2O and cupric oxide CuO, made by heating copper in air. It is a red , brown or black powder.

Aes viride : Another name for verdigris, 'Gruenspan' in German.

Aestas,		Sommer.
Ahenum.		Ein kupferner oder eiserner Keſſel.
Albumen.		Das Eyerweis.
Alcali, alkali ſal, ſ. Cineres clavellati, und Sal alcali.		Aſchen=Salz, ſiehe alumen catinum.
Alcohol vini, Spiritus vini rectificatiſſimus.		Der allerſtärckſte Brandewein.
Alembicus.		Deſtillir=Helm.
Alumen.		Alaun.
Alumen calcinatum, uſtum.		Gebrannter Alaun.
Alumen catinum.		Potaſche, Weidaſche.

4

Aestas : Summer.

Ahenum : A copper vessel used for a water bath. It has a lid with a hole for the head of a flask to exit.

Albumen : Egg white.

Alcali, alkali sal : See Cineres clavellati and Sal alcali. Potash, potassium carbonate K_2CO_3, made by extraction from the ashes from burning wood.

Alcohol vini, Spiritus vini rectificatissimus : Alcohol made by distilling from wine.

Alembicus : An alembic. A distillation apparatus. Usually it is a glass vessel, set on top of a flask to collect distilled vapours and allow them to be collected in a receiving vessel. Often an alembic has a characteristic beak which fits the receiving flask and a helmet or head which fits the flask in which the substances are being heated.

Alumen : Alum. Potassium aluminium sulphate $KAl(SO_4)_2 \cdot 12H_2O$. Found in various minerals. As it was used in dyeing, it was a part of commerce from Greek times.

Alumen calcinatum, ustum : Burnt alum. On heating alum loses the water bound up within it and becomes anhydrous.

Alumen catinum : Another name for potash. Not connected with the substance alum.

Alumen plumeum, plumofum. — Federweis.

Alumen facchari-num, zaccarinum. — Zucker=Alaun.

Alumen uftum. — Gebrannter Alaun.

Amalgama. — Amalgama.

Ammoniacum, fiehe Sal ammoniacum. — Salmiack.

Ana. — Jedes gleichviel.

Animalia. — Die Thiere.

Antimonii flores. — Spiesglas=Blüth.

Antimonii hepar. — Spiesglas=Leber.

Antimonii regulus. — Spiesglas=König.

6

Alumen plumeum, plumosum : Feathered or hairy alum. $Al_2(SO_4)_3 \cdot 17H_2O$. This is in the form of fine needle like crystals.

Alumen saccharinum : A confection made from alum mixed with eggwhite and rose water, and baked into small loaves.

Alumen ustum : Burnt alum.

Amalgama : An alloy of a metal with mercury.

Ammoniacum see **Sal ammoniac** : Ammonium chloride NH_4Cl. A mineral found around volcanic vents deposited through sublimation.

Ana : Each in the same quantity. Add the same quantity.

Animalia : Any substance of animal origin.

Antimonii flores : Flowers of antimony. Antimony trioxide Sb_2O_3. Found as a mineral and is purified by sublimation when it forms a white powder. It can also be made by heating antimony metal in a current of air and subliming. First described in the writings of Basil Valentine.

Antimonii hepar : Liver of Antimony. A substance, of a liver-brown colour, made by fusing antimony sulphide with sulphur and salt.

Antimony regulus : Metallic antimony, Sb.

Antimonii vitrum.  Spiesglas-Glanz.

Antimonium, siehe
antimonium spa-
gyrice praepar:

Antimonium spagy-
riae praeparatum. Spiesglas.
Spiesglanz.

Aphronitrum, siehe
Sal petrae.

Aqua. Waffer.

Aqua fontana. Font, — — —Brunnen-Waffer.

Aqua fortis sim-
plex, aqua gehen-
nac, stygia.

8

Antimonii vitrum : Antimony glass. Made by fusing mixtures of antimony trioxide and trisulphide. It was called Spiessglanzglas, antimony glass, or vitrum antimonii.

Antimonium : See antimonium spagyice praepar.

Antimonium spagyriae praeparatum : Antimony prepared spagyrically. Appears to be related to Spiessglas or glass of antimony.

Aphronitrum : See Sal petrae.

Aqua : Water.

Aqua fontana : Spring water.

Aqua fortis simplex, aqua gehennae, stygia : Strong water, or concentrated Nitric acid, HNO_3.

Aqua pluuialis.	▽ ♇ R, ⊽ ▽ R, ▽	Regen=Waſſer.
Aqua regis.	▽ ▽ A, ▽, ≈≈R ♃ ▽ ⅄, ≈≈ ⅋,	Goldſcheid=Waſſer.
Aqua vitae.	♍ ∜ 8, ⊞ ⊹, ⇆, A⌿ ⊕ T, I, ▰▰ ⊽, ∴ ℞, ♇ ∜ ∜, A⌞, N°°℈, Ψito ⚸ ⊟,	Aquavit , Lebens= Waſſer.
Arena.	△ ℔ ∴, S, ≣ ≈ –, ⊓, ✝ ⣿, ⣿,	Sand.
Argentum, luna.	☽, ⊂⊽⊃ ℂ, ⊕, ℰ ∿ ∾, ♀ ℀ ⋔ ∀ ℳ, ⸫ W, ☾,	Silber.

10

Aqua pluvialis : Rain water.

Aqua regis : Royal water, Aqua regia. Mixture of concentrated nitric HNO_3 and hydrochloric HCl acids. This is able to dissolve gold. Apparently was first described in a 14th century work of pseudo-Geber.

Aqua vitae : Water of life. Usually applied to alcohol C_2H_5OH distilled from wine.

Arena : Sand.

Argentum, luna : Silver.

Argentum folia-
tum.
                              Silber = Blättlein.

Argentum muſi-
cum.
                              Saiten = Silber.

Argentum picto-
rium.
                              Mahler = Silber.

Argentum viuum,
Mercurius viuus,
Hydrargyrum.
                              Queck = Silber.

Armena bolus.
                              Armeniſcher Bolus.

Arſenicum album.
                              Weiſſer   Arſenic,
                              Maus = Gift, Rat=
                              ten = Gift.

Arſenicum citri-
num, flauum lu-
teum.
                              Rauſchgelb.

Argentum foliatum : Silver leaf.

Argentum musicum : Mosaic silver. Presumably a parallel to mosiac gold, a compound of tin used to gild metals.

Argentum pictorium : Silver paint.

Argentum vivum, **mercurius vivus**, **Hydrargyrum** : Metallic mercury.

Armena bolus : An earthy clay, native to Armenia. It is usually red in colour red due to the presence of iron oxide. The clay also contains hydrous silicates of aluminum and possibly magnesium.

Arsenicum album : White arsenic. Arsenic trioxide As_2O_3, a white powder.

Arsenicum citrinum, **flavum luteum** : Yellow to golden-coloured, flaky mineral containing arsenic in the form of arsenic sulphide As_2S_3, orpiment.

Arſenicum rubrum, Sandaracha grae- corum. Rauſchgelb, rother Operment.

Arſenicum ſublima- tum. Sublimirter Arſe- nick.

Atramentum, Vi- triolum. Dinte, Vitriol.

Atramentum al- bum, Vitriolum album, Kupfer = Waſſer, weiſſer Vitriol.

Aurichalcum, cu prum citrinum. Meſſing.

Auripigmentum, Riſigallum. Operment.

Arsenicum rubeum, Sandaracha graecorum : Red arsenic, the Sandarach of the Greeks. Also known as ruby sulphur or ruby of arsenic. Red form of arsenic sulphide As_4S_4 Realgar.

Arsenicum sublimatum : Arsenic sulphide or orpiment As_2S_3 sublimes readily.

Atramentum, Vitriolum : Iron (Ferrous) Sulphate $FeSO_4$.

Atramentum album, Vitriolum album : Zinc Sulphate $ZnSO_4$.

Aurichalcum, cuprum citrinum : A name applied to gold coloured alloys of copper, similar to brass, a copper tin alloy. Some alchemists reserved this for gold copper alloys.

Auripigmentum, Risigallum : Orpiment.

Aurum, Sol.	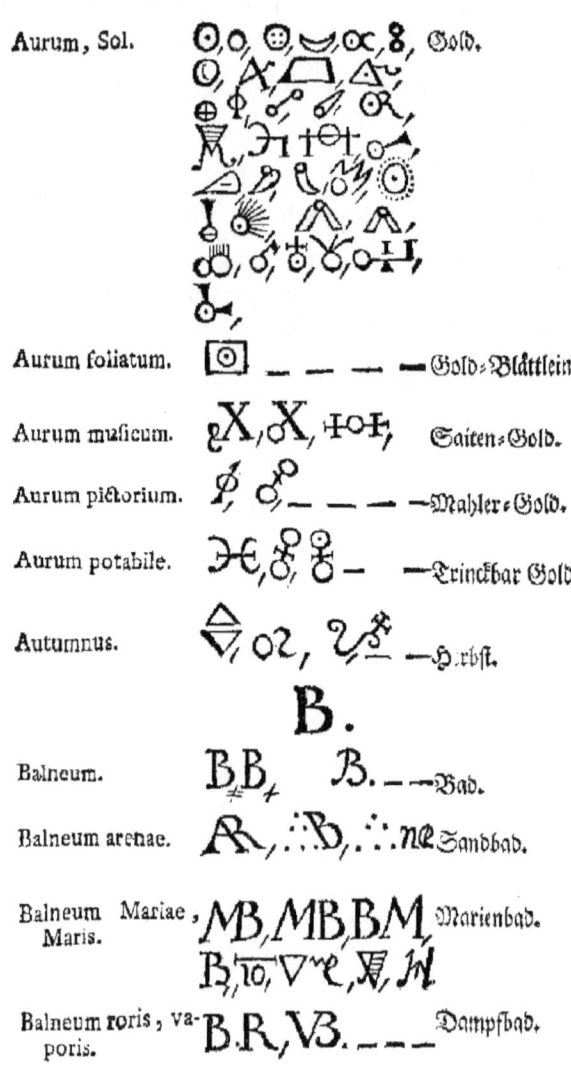 Gold.
Aurum foliatum.	⊡ — — — — Gold-Blättlein.
Aurum muſicum.	℞X, ₰X, ⊹○⊢, Saiten-Gold.
Aurum pictorium.	∅, ℰ° — — — — Mahler-Gold.
Aurum potabile.	♓︎♄, ℰ°, ℥ — — Trinckbar Gold.
Autumnus.	⬙, ♌︎, ♌ⅈ — Herbſt.

B.

Balneum.	B,B, ℬ. — — Bad.
Balneum arenae.	℟, ∴B, ∴ℳ Sandbad.
Balneum Mariae, Maris.	MB, MB, BM, Marienbad. B, ⅉ, ∇ℓ, ℥, ℋ
Balneum roris, va- poris.	B.R, V3. — — — Dampfbad.

16

Aurum, Sol : Gold.

Aurum foliatum : Gold leaf.

Aurum musicum : Aurum musivum or mosiacum, mosaic gold, Stannic sulphide SnS_2 is made by fusing tin and sulphur.

Aurum pictorium : A gold coloured pigment used by painters and in illuminated manuscripts to mimic gold. One recipe for this malergold was a preparation of tin, sal ammoniac, sulphur and mercury. In other cases the name was applied to mosaic gold.

Aurum potabile : Gold made into a soluble or drinkable form.

Autumnus : Autumn.

Balneum : A bath or apparatus for the controlled heating of a flask.

Balneum arenae : Sand bath or tray of sand heated from below, thus ensuring a steady heating of a flask placed in the sand.

Balneum Mariae, Maris : Water bath. A vessel of water heated from below into which the flask was placed. The maximum temperature that could be obtained was thus the boiling point of water.

Balneum roris, vaporis : Steam or vapour bath. A vessel of water heated from below, but here the flask was suspended above the water in the steam. This could achieve temperatures a little above that of boiling water.

Benzoe flores, ſiehe
Flores benzoe.

Bezoar occidenta-
lis.   ⬭ — — — — — Weſt-Indiſcher Bezoar.

Bezoar orientalis.   ⊙ — — — — — Oſt-Indiſcher Bezoar.

Bezoardicum Jo-
uiale.   ♃ — — — — — Schweistreibend
Zinn.

Bezoardicum luna-
re.   ☽ — — — — — Schweistreibend
Silber.

Bezoardicum mar-
tiale.   ♂ — — — — — Schweistreibend
Eiſen.

Bezoardicum mine
rale.   ☿ — — — — — Schweistreibender
Spiesglas-Kö-
nig.

Bezoardicum Satur-
ninum.   ♄ — — — — — Schweistreibend
Bley.

Bezoardicum ſola-
re.   ☉ — — — — — Schweistreibend
Gold.

Bezoardicum vene-
reum.   ♀ — — — — — Schweistreibend
Kupfer.

Bismuthum Marca-
ſita.   ⧆ , 2ſ — — Wismuth.

Bolus alba.   ℔ — — — — — Weiſſer Bolus.

Bolus armena, ſiehe
armena bolus.

Bolus communis.   d , ♌ , ℅ — — Gemeiner Bolus.

Benzoe flores : see Flores benzoe.

Bezoar occidentalis : Bezoar stones from the stomach or galls of Western animals.

Bezoar orientalis : Bezoar stones from the stomach or galls of Eastern animals.

Bezoardicum Joviale : Preparation of butter of antimony and a solution of tin in nitric acid, heated to melting point. Its main constituent is antimony tetroxide Sb_2O_4 and tin dioxide SnO_2 .

Bezoardicum lunare : Preparation of butter of antimony and a solution of Silver in nitric acid, heated to melting point. Its main constituent is antimony tetroxide Sb_2O_4 and silver chloride $AgCl$.

Bezoardicum martiale : Preparation of butter of antimony and a solution of iron in nitric acid, heated to melting point. Its main constituent is antimony tetroxide Sb_2O_4 and Iron (Ferric) oxide Fe_2O_3 .

Bezoardicum minerale : Preparation of butter of antimony and alcohol in nitric acid, heated to melting point. Its main constituent is antimony tetroxide Sb_2O_4 .

Bezoardicum Saturninum : Preparation of butter of antimony and lead in nitric acid, heated to melting point. Its main constituent is antimony tetroxide Sb_2O_4 and lead chloride $PbCl_2$.

Bezoardicum solare : Preparation of butter of antimony and a solution of gold in aqua regia, heated to melting point. Its main constituent is antimony tetroxide Sb_2O_4 and finely divided gold.

Bezoardicum venereum : Preparation of butter of antimony and a solution of copper in nitric acid, heated to melting point. Its main constituent is antimony tetroxide Sb_2O_4 and copper oxide CuO.

Bismuthum Marcasita : Known to the German miners of the 16th century as Wismuth or "white mass". Sometimes called "tin-glass". The marcasite of Bismuth is likely to be Bismuth sulphide Bi_2S_3.

Bolus alba : White clay, sometimes specifically applied to kaolin or china clay.

Bolus armena : See **Armena bolus**.

Bolus communis : Common potter's clay.

Borax, Borrax.

Borax.

C.

Cadmia factitia, for-
nacum, siehe

Tutia, Tutien,
Tutz.

Cadmia fossilis, pa-
tiua, lapis calami-
naris.

— — — — Gallmey-Stein.

Calcinare,

Rösten, ausglühen,
in ein Pulver ver-
brennen.

Calcinatio argenti.

Das Ausglüen des
Silbers in ein
Pulver.

Calcinatio auri.

Das Ausglüen des
Golds in ein Pul-
ver.

Calx.

Kalch von Metallen.

Calx ouorum.

Calcinirte Eyerscha-
len.

Calx Solis.

— — — — Gold-Kalch.

Calx viua.

Ungelöschter Kalch.

Borax, Borrax : Sodium tertraborate $Na_2B_4O_7$, found as evaporation deposits around salt lakes. It was widely used in the 15th and 16th centuries as a mordant in dyeing, a soap and also as a flux for soldering metals.

Cadmia factitia, fornacum : Oxides of zinc, primarily ZnO, which sublimes onto the cooler sides of furnaces or chimneys, in which copper or brass was smelted. Also called "tutty".

Cadmia fossilis, pativa, lapis calaminaris : Calamine, a mineral substance of greyish, brownish, yellowish, or pale reddish colour. It is primarily compounded from Zinc oxide ZnO with some Zinc carbonate $ZnCO_3$ and Ferric oxide Fe_2O_3.

Calcinare : To calcine or burn at a high temperature, sometimes reducing a substance to an ash.

Calcinatio argenti : Silver calcined with common salt $NaCl$ forms Silver chloride $AgCl$.

Calcinatio auri : Gold was sometimes calcined with mercury and sal ammoniac. The chlorine from the sal ammoniac probably formed auric chloride Au_2Cl_6, a dark red powder.

Calx : Any powder or ash formed by calcining or roasting a substance.

Calx ovorum : Calcined eggshells, primarily calcium carbonate, $CaCO_3$. When finely powdered could be used as an artist's pigment.

Calx Solis : Calcined gold.

Calx viva : Quicklime or calcium oxide CaO, formed by roasting limestone.

Camphora.	~~, ~~~, ~~~~, Campher.
Cancer, aftacus, Gammarus.	$\mathfrak{S}$ – – – – – Ein Krebs.
Capella.	$\mathcal{T}\,\mathcal{P}\,O\,\mathcal{X}\,\mathcal{X}$, Sand=Capelle, Capelle.
Caput mortuum.	Todtenkopf.
Catinus, tigillum.	– – – – – Tiegel.
Caementare, ftratificare.	Cämentiren.
Cera citrina.	Gelb Wachs.
Cerufla, plumbago, plumbum album.	Bleyweis.
Chalybs, ferrum.	Stahl, Eisen.
Cineres clauellati, f. auch alkali fal.	Pottafche.

22

Camphora : A flammable, white or transparent resin with a strong aromatic odour found in the wood of the camphor laurel (Cinnamomum camphora).

Cancer, astacus Gammarus : The astrological sign of Cancer, the crab.

Capella : Hair.

Caput mortuum : Death's head, or residue in the crucible after calcining a substance to a very high temperature usually in a crucible.

Catinus, tigillum : A small bowl sometimes made from bone ash for purifying silver or gold by blowing air across it.

Caementare, stratificare : To cement or stratify, is to heat layers of different substances to a very high temperature usually in a crucible.

Cera citrina : Beeswax.

Cerussa, plumbago, plumbum album : White lead or lead carbonate $PbCO_3$.

Chalybs, ferrum : Steel or iron.

Cineres, clavellati or alkali sal : Potash, derived from wood ashes, contains potassium carbonate, K_2CO_3 and potassium hydroxide KOH.

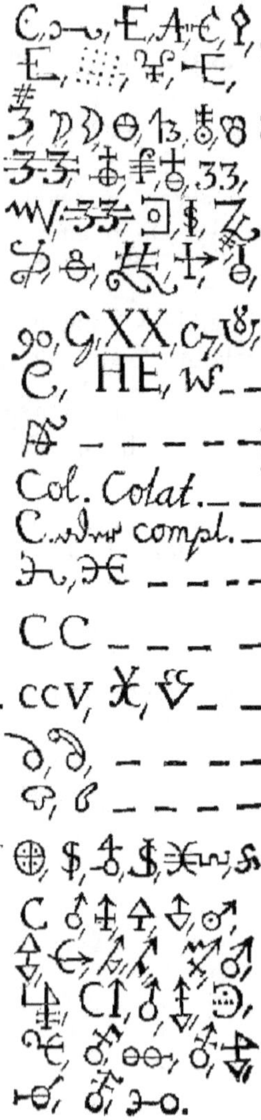

Cinis , cineres.	C,⸜⸝, E, A, ⊖, ♀, E, ⁘, ⅋, E,	Aſche.
Cinnabaris.	♯, ♄, ☽, ☽, ⊕, ⅓, ⚡, ⚥, ⅗, ⅗, ⚚, ⚡, ⚡, 33, ♏, ⅗⅗, ⊡, ⚡, ⚡, ♄, ⚥, ⚡, ✝, ⚡,	Zinnober.
Coagulatio.	90, G, XX, C⁊, ♉, C, HE, ♏ _ _	Coaguliren.
Cobaltum.	♉ _ _ _ _ _ _	Mucken-Gift.
Colaturæ.	Col. Colat. _ _	Das, was durchge= ſeyht, oder durch=
Completus.	C.⸝⸍ compl. _	geſiegen iſt. Das Complete.
Compoſitio.	♃, ♓ _ _ _ _	Vermiſchung, Zu= ſammenſetzung vie=
Cornu cerui.	C C _ _ _ _	ler Artzneyen. Hirſchhorn.
Cornu cerui uſtum.	CCV, ♉, ♉ _ _	Gebrannt Hirſch= horn.
Cornuta.	☽, ☽, _ _ _ _ _	Retorte.
Creta.	♎, ℓ _ _ _ _ _	Kreide.
Crocus, crocus aro- maticus.	⊕, ♃, ⚡, ♃, ♓⸜⸝, ♄	Saffran.
Crocus martis.	C, ⚡, ⚡, ⚡, ⚡, ♂ ♃, ⚡, ⚡, ⚡, ⚡ ♃, CT, ⚡, ⚡, ☽, ⚡, ⚡, ⊖⊖, ⚡, ♃ ⚡, ⚡, ⊋⊙.	Eiſen=Saffran.

Cinis, cineres : Ashes.

Cinnabaris : Cinnabar, Mercuric sulphide HgS, the most common ore of mercury. When purified and ground to a powder is used as the red pigment vermilion.

Coagulatio : Coagulation, or making a solution thicken and clump together.

Cobaltum : Ore of cobalt. Ruland states "is a Metallic substance, darker than lead or iron, sometimes of ashen hue, wanting in metallic tint or lustre. It can be melted and made into plates. It is not fixed, but takes away the higher metals with it in smoke."

Colaturae : A substance that has been purified or clarified through filtration.

Completus : A process which is completed or satiated.

Compositio : To mix or compose substances together.

Cornu cervi : Deer or stag horn.

Cornu cervi ustum : Burnt deer or stag horn.

Cornuta : A retort used in distilling.

Creta : Chalk or Calcium carbonate $CaCO_3$.

Crocus, crocus aromaticus : Saffron red-yellow spice made from crocus flowers.

Crocus martis : Red rust or oxide of Iron, FeO.

Crocus metallo- rum.		Spiesglas = Saff= ran.
Crocus veneris.		Kupfer = Saffran.
Crucibulum, f. Ca tinus, und Tigil- lum.		Schmeltz = Tiegel.
Cryftallus.		Cryftall.
Cucurbita.		Ein gläferner Kolbe.
Cucurbita cœca. Cuprum, Venus, fie= he aes.		Ein blinder gefchlof= fener Kolbe.
Cum vino.		Mit Wein.

D.

Da & figna, oder de- tur, fignetur.		Gib und überfchrei= be es.
Deftillare, deftilla- tio.		Brennen, deftilliren.
Dies.		Tag.
Dies & nox, nyct- hemeron.		Tag und Nacht.

Crocus metallorum : Red brown liver of antimony, powerfully emetic. A liver-brown substance made by fusing from antimony sulphide and metal sulphides resulting in a complex mixture of sulphides, from which crocus metallorum was extracted with water.

Crocus veneris : Red oxide of copper, cupric oxide Cu_2O

Crucibulum : See Catinus and Tigillum.

Crystallus : Crystal.

Cucurbita : A flask. Often used for the flask section of an alembic, or still, to which a still head was fixed.

Cucurbita coeca : Blind cucurbit, made from pottery rather than glass.

Cuprum Venus : See aes.

Cum vino : With wine.

Da and signa, or detur, signetur : Explained in writing.

Destillare, destillatio : Distillation. Distillate.

Dies : A day in length.

Dies et nox, nycthemeron : Period of a day and a night.

Digerere, digeftio.	$\mathcal{D}\mathcal{G}, \rightleftharpoons, \overline{33}, \overline{8},$ $\overline{8}, 48, \mathcal{D}.$	Digeriren, erwär= men.
Drachma, Holca.	$3, \triangle$ _ _ _	Drachme, Quent= lein; der vierte Theil eines Loths, 6o. Gran, 3. Scrupel.
Drachma femis.	3β _ _ .. _	Ein halb Quentlein.

E.

Ebullitio.	A _ _ _ _	Das Braufen, Sie= den.
Elementa, principia corporum.	$\bigoplus, \amalg, \mathcal{A}, \bigstar,$	Die Grund=Theile der Cörper.
Effentia.	$\maltese$ _ _ _ _	Eine Effenz.
Excipulum, fiehe Re- ceptaculum.		
Extractio ficca.		Siehe Sublimatio, das Sublimiren.

F.

Farina.	$\odot$ _ _ _ _	ein Mehl, feines Pulver.
Farina laterum.	$\#, \boxdot, \boxdot, \boxdot, \boxdot$	Ziegel=Mehl.
Faex, faex vini, ace- ti.	$\mathcal{H}, \ni\in$ _ _ _	Wein= oder Effig= Hefen.
Fel vitri, Sal vitri.	$\mathcal{b}, \Omega, \square, \mathcal{S}$	Glas=Galle.

Digerere, digestio : Digestion or the breaking down of a substance usually in some liquid.

Drachma, Holca : An eighth of an ounce in weight. About four grams in modern units.

Drachma semis : Half a drachma in weight.

Ebullitio : To boil.

Elementa, principia corporum : The elements, basic parts, or foundation principles of a body.

Essentia : Essence.

Excipulum : See Reciptaculum.

Extractio sicca : Dry extraction, or sublimation.

Farina : Flour.

Farina laterum : Powdered brick.

Faex, faex vini, aceti : Dregs of wine, vinegar.

Fel vitri, Sal vitri : Glass gall, sandiver. A liquid saline matter found floating over glass after vitrification.

Ferrugo, ferri vitium, situs. ‡‡, ♃‡ — — — Eisen-Rost.

Ferrum, siehe Mars. ♂, ⇌, ⚔, ☿, ♌, F, ⚔, L, ♉, Eisen.

Filtratio, philtratio. ⊓, ☊, ☿, ⚹, 33 ♀ Das Filtriren durchseihen, durch ein Lösch-Papier.

Filtrum philtrum. ℔ — — — — — Ein Filtrir-Glas mit dem Zugehör.

Fimus equinus. ℃, Ω, ⚏, ⚒ Pferd-Mist oder andere feuchte Wärme von Asche oder warmen Wasser zu dem Digeriren.

Fixus, fixum. V — — — — — Feuerbeständig.

Figere, fixatio. V, ♈, ⚏, ♄ — . Figiren, etwas Flüchtiges Feuerbeständig machen.

Flores. Fl. — — — — Blumen.

Flores antimonii, siehe antimonii flores. Spiesglas-Blumen.

- - benzoe. F — — — — Benzoin-Blumen.

- - Martis, siehe crocus Martis. Stahl-Blumen Stahl-Saffror

30

Ferrugo, ferri vitium, fitus : Iron rust.

Ferrum : Iron. see also Mars.

Filtratio, philtratio : Filtration.

Filtrum, philtrum : Material which has been filtered.

Fimus equinus : Horse dung, often used to make a bath that generates a slow mild heat.

Fixus, fixum : Fixed.

Figere, fixatio : To fix.

Flores : Flowers. Often used figuratively for powdery sublimation products such as flowers of sulphur.

Flores antimonii : See antimonii flores.

Flores benzoe : Flowers or sublimate of the resin storax exuded from the Benjamin tree.

Flores Martis : See crocus Martis.

Flores vitroli.	⅃‾ — — —	—Vitriol-Blumen.
Flores viridis aeris.	℞, ⚇ — — —	—Grünspan-Blumen.
Fluere.	⊸∿, ∘—ℨ,∿℄,₣,	Fliessen.
Fornax, furnus.	⊡,-⊖,⊡ — —	Ein Ofen.
Fuligo.	⌀, ⅂⅂,♌,✠,	Ruß.
Fumus.	⊡, ℓ, ∿ — —	Rauch.
Furnus, siehe Fornax.		
Fusio.	⹀Ey — — —	—Das Schmelzen.

G.

Gummi.	G,ℐℓℨ,⅜,⌽⌽⌀,∘-ᴄ ℐℓℐ, ℓℓℐ,	Ein Gummi, Hartz.
Gummi arabicum.	⹀Ⅎ — — — —	Arabischer Gummi.
Gradatio.	Mav,ℑ,Mva	Das Gradiren, Erhöhen der Metalle.
Gradus ignus.	⌢₂ — — — —	—Grad des Feuers.
Granatus.	⊕⹀ — — — —	Granat-Stein.
Granum.	gr. — — — — —	Ein Gran, der 60ste Theil eines Quintleins, oder 20ste einer Scrupel. Gut-

Flores vitrioli : Flowers or dehydrated copper sulphate $CuSO_4$. Copper sulphate is in the form of blue crystals but when heated it turns into a white powder.

Flores viridis aeris : Flowers or sublimate of verdigris, copper acetate $Cu(CH_3CO_2)_2$.

Fluere : To flow, run down or drip.

Fornax, furnus : Furnace.

Fuligo : Soot.

Fumus : Fumes, vapours or smoke.

Furnus : See Fornax.

Fusio : To fuse or melt.

Gummi : Gum.

Gummi arabicum : Gum arabic, made from the hardened sap of the acacia tree.

Gradatio : The gradual purification of a substance, often through a series of stages

Gradus ignus : Various grades or degrees of fire.

Granatus : Seeds or grains.

Granum : A grain. A small quantity. There are 480 grains in an ounce, or 60 in a drachma.

Gutta, guttae.	*G. g. gtt* _ _ _	Ein Tropfen; Tropfen.

H.

Haematites , siehe Lapis haematites.		
Herba.	*H, HB* _ _ _	Ein Kraut.
Hermetice sigillatum.	*H. S.* _ _ _ _	Hermetisch sigillirt, zugeschmeltzt.

Hora.	⧖, ℩~, *A, Ⅱ,* Γ, ⴱ, *H,* ⧣, ⧖, ♉	Eine Stunde.
Hiems.	Ⅲ, ♒ _	— Der Winter.
Hydrargyrum, siehe Argentum vivum.		

I.

Ignis.	△, △, □, Z, §§, 20	Das Feuer.
Ignis circulatorius.	△, ⊿, ▣, λ _ _	Ein gelind, Circulir-Feuer.
Ignis fortis.	∽△, △△ _ _ _	Ein starck Feuer.
Ignis lentus.	♇, △, △̄, ⌞△,	Ein langsam, gelind Feuer.
Ignis reuerberius.	△R _ _ _ _	Ein Reverberir-Feuer.
Ignis rotae.	⊘ _ _ _ _ _	Ein Rad-Feuer.
Imbibere.	✡ _ _ _ _ _	Träncken.

Gutta, guttae : A drop, drops.

Haematites : See Lapis haematites.

Hermetice sigillatum : Hermetically sealed.

Hora : An hour.

Hiems : Winter.

Hydrargyrum : See Argentum vivum.

Ignis : Fire.

Ignis circulatorius : Circulating fire. A moderate fire used to heat substances in a circulating or reflux vessel, such as a pelican where the distillate was continuously returned to the heated flask.

Ignis fortis : A strong fire.

Ignis lentus : A gentle fire.

Ignis reverberius : Reverberating fire, a very strong fire usually created using bellows. In a reverberating furnace the substance was usually in direct contact with the flames, rather than being enclosed in a crucible.

Ignis rotae : A strong fire for melting materials in a crucible. The wood or coals were placed around the crucible, thus forming a circle or wheel (rota) around the crucible.

Imbibere : To drink. The adding of a liquid to the materials reacting in a flask.

| Incompletus. | *Inc. Pincompl.* | Das Incomplete. |

Jupiter, ſtannum. ... Zinn.

L.

Lege artis. $\mathcal{L}$ *a.* ... *l. a.* Nach der Kunſt.

Lamina.   — — — Ein Blech.

Lana illota, Erion.   — — — Ungewaſchene Wolle.

Lapides.   — — — Steine.

Lapis armenius, armenus, Malachites.   — — — Armenien = Stein, Bergblau.

Lapis bezoar occidentalis, ſ. Bezoar occidental.

- - - - Orientalis, ſ. Bezoar orientalis.

- - calaminaris, ſ. Calaminaris.   Gallmey = Stein.

- - calcarius.   Kalk = Stein.

- - haematites.   Blut = Stein.

- - Lazuli.   Laſur Stein.

Incompletus : Incomplete.

Jupiter, stannum : Tin, associated with the planet Jupiter.

Lege artis : The laws of the art.

Lamina : Thin slices, plates, sheets or leaves. Often metals were beaten into thin plates to make them react quicker in some process.

Lana illota, Erion : Unwashed wool. "Erion" is Greek for "wool".

Lapides : Stones.

Lapis armenius, armenus, Malachites : Blue ochre or Armenian stone. A sky or pale blue copper mineral found in Germany and Hungary, often in silver mines. It was used as artists pigment sometimes replacing the expensive lapis lazuli.

Lapis bezoar occidentalis : See Bezoar occidental.

Lapis bezoar orientalis : See Bezoar orientalis.

Lapis calaminaris, Calaminatis : The main ore of zinc, primarily Zinc oxide ZnO.

Lapis calcarius : Limestone, primarily Calcium carbonate $CaCO_3$.

Lapis haematites : Bloodstone, a hard, heavy, dark-red mineral, mainly composed of iron oxide Fe_2O_3.

Lapis Lazuli : The expensive blue pigment, obtained by grinding a mineral found only in Asia, primarily Afghanistan. Thus it was known as "ultramarine" coming from beyond the seas.

Lapis Magnes , Sideritis Plinii, Lapis nauticus.	♂, ♂♂, ⊹, ♁,	Magnet - Stein , Magnet, Segel-Stein.
Lapis fabulofus, o-ſteocolla.	⌇⌇⌇⌇	— — —Bein-Bruch.
Lapis filex.	° ° °	— — — — —Kiefel, Kiefel-Stein.
Later.	⊞, ▭, ▨	— —Ziegel-Stein.
Lateres cribrati.	⊞, ⊡⊡, ▣, ▱	Geſiebte Ziegel-Steine.
Libra:	♎ ℔ ♏ ℔, ⚖ ♐ ℔24℔℔	Ein Pfund.
Libra ciuilis , pondus ciuile.	C. ρ.	— — — —Ein gemein Pfund von 32. Loth.
Libra medicinalis.	ℳ. ρ.	— — — Ein Apothecker-Pfund von 24. Loth oder 12. Onzen.
Libra penſilis.	⚖	— — — —Eine Wage.
Lignum.	♄, ♄	— — —Holz.
Limatura chalybis, martis.	♂, ⊙⟶, ♂	— Feil-Staub , Eiſenfeil.
Lixiuium, ſiehe auch alcali.	♃, ♏, ♏, ♌♂, △, ♇	Eine Lauge , ein Laugen-Salz.

38

Lapis Magnes, Sideritis Plinii, Lapis nauticus : Lodestone, a hard, heavy, magnetic ore of iron, primarily magnetite, iron oxide Fe_3O_4, often bluish or brown in colour.

Lapis sabulosus, osteocolla : A petrified tree root, or stalactite, of Calcium carbonate $CaCO_3$. A friable chalky substance, sometimes called "bone-binding stone" used medically to treat broken bones. "Sabulosus" means sandy or gravelly.

Lapis silex : Flint.

Later : Brick.

Lateres cribrati : Brick dust, ground fine enough to pass through a sieve.

Libra : Scales, or constellation of the Scales.

Libra civilis, pondus civile : A civil scale for weighing quantities used in commerce. A pound consists of 32 loth.

Libra medicinalis : Medicinal scales, for weighing the small quantities necessary in medicine. Apothecary scales and weights. A pound consists of 24 loth or 12 ounces.

Libra pensilis : A set of scales in which the weights are hung from a thread.

Lignum : Wood.

Limatura chalybis, martis : Iron filings.

Lixivium : Lye or alkali. Usually applied to Potash, potassium carbonate K_2CO_3, made by extraction from the ashes from burning wood.

Luna, ſiehe Argen-
  tum.

Lutatio.
                 Silber.
                 Das   Verlutiren,
                 Verklaiben   der
                 Gefäſſe.

Lutum.
                   Ein Leim oder Kütt.

Lutum Philoſopho-
  rum, lutum ſa-
  pientiae.
                 Der   philoſophiſche
                 Leim oder Kütt.

Magnes, ſiehe Lapis
  magnes.

Manipulus , Man-
  nes.
                 Eine Handvoll.

Magneſia.
                 Magneſien.

Marcaſita, ſiehe Bis-
  muthúm.
                 Marcaſit , Wis-
                 muth.

Marcaſita aurea,
  oder metallica, ſihe
  Zincum.
                 Zinck.

Mars, ſiehe Ferrum.
                 Eiſen.

Luna : Silver. See Argentum.

Lutatio : To lute or seal up gaps between a flask and a still head, or the joint with a receiver.

Lutum : Lute or cement for sealing the joints between vessels. Usually a mixture of clay and flour made into a paste with fibres of hair or grasses added to give it strength.

Lutum Philosophorum, lutum sapientae : The lute of the philosophers. Sometimes refers to a coating of lute applied to glass vessels to help prevent cracking.

Magnes : See Lapis magnes.

Manipulus, Mannes : A handful.

Magnesia : A name applied to many different substances. Sometimes confused with magnes. Ruland says Magnesia generally stands for Marcasite and that artificial magnesia is melted tin into which mercury has been injected, and the two have been mingled together until they form a brittle substance, and a white mass. He also holds that it is silver mixed with mercury, forming an extremely fusible metallic compound which is liquefied as easily as wax, is of a wonderful whiteness, and is called the magnesia of the philosophers. In the later period of alchemy, it became applied to white minerals primarily containing magnesium compounds. The name comes from the Magnesia region in Greece where magnesia alba, the ore of Magnesium oxide MgO, was found.

Marcasita : Bismuth or tin-glass, found in tin mines. Metallic bismuth Bi was known from the beginning of the 15th Century, but often confused with other metals such as zinc and antimony.

Marcasita aurea, or metallica : Zinc. See also Zincum.

Mars : Iron. See also Ferrum.

Maſſa.	[symbols]	Eine Maſſe oder ein Taig zu Pflaſtern, Pillen.
Maſſa pilularum.	[symbol] — — — —	Eine Pillen=Maſſe.
Materia.	$\hat{a}\,\hat{a}$, maa — —	Eine Materie.
Materia prima.	[symbols]	Eine Grund=Mate= rie.
Mel.	[symbols]	Honig.
Menſis.	[symbols]	Ein Monat.
Mercurius vinus, ſ. Hydrargyrum, argentum viuum.	[symbol]	
Mercurius praeci- pitatus albus.	[symbols]	Weiſſer Präcipitat von Queck=Sil= ber.
Mercurius praeci- pitatus ruber.	[symbols] — — —	Rother Präcipitat.
Mercurius Saturni praecipitatus, Mi- nium.	[symbols]	Mening, Mini.
Mercurius sublima- tus.	[symbols]	Sublimat.

Massa : A quantity of substance that could be made into a pill.

Massa pilularum : A pill.

Materia : Matter.

Materia prima : The first matter, or the substance from which one starts the alchemical process.

Mel : Honey.

Mensis : A month.

Mercurius vivus : See Hydrargyrum and argentum vivum.

Mercurius praecipatus albus : Mercurous chloride Hg_2Cl_2, a dense white or yellowish-white powder. Also known as corrosive sublimate as it sublimes at a relatively low temperature. Can be made from soluble mercurous salts by precipitation using common salt.

Mercurius praecipatus ruber : Mercuric oxide HgO, a red or orange coloured solid. Can be made by precipitation from soluble mercuric salts using alkalis.

Mercurius Saturni praecipatus, Minium : Red lead, or Triplumbic tetroxide, PB3O4. Probably named because it is made by heating molten lead, appearing like liquid mercury, in a current of air, and thus this mercury of Saturn or lead appeared to precipitate the red oxide.

Mercurius sublimatus : Mercurous chloride Hg_2Cl_2, a dense white or yellowish-white powder known as corrosive sublimate as it sublimes at relatively low temperature.

Minium, fiehe Mer-
   curius   Saturni
   praecipitatus.
                                 Mening, Mini.

Mifce , NB. am En- *M.* — — —   Mifche.
   be ber Recepte.

Mixtura   fimplex *M. S L*   —D. Ludwigs fimple
   Ludouici.   Tropfen.

N.

Numero.   *Nr. N̲o̲* — — An der Zahl. Wird
                                  z. E. gebraucht,
                                  wo man Früchte
                                  verfchreibt.

Nitrum commune.   Salpeter.

Nox.   Eine Nacht.

Nux mofchata. *M* — — — —Mufcatnus.
Nycthemeron, fiehe
   dies & nox.   Tag und Nacht.

O.

Obulus   fcrupulus *Ʒß*   Ein halber Scrupel
   femis.   — — — — oder 10. Gran.

Minium : See Mercurius Saturni praecipitatus.

Misce : Mix.

Mixtura simplex Ludovici : A medicine consisting of Tincture Bezoarde, Spiritus Tartari, Vitriol and Camphor.

Numero : Number.

Nitrum commune : Saltpetre or Potassium Nitrate KNO_3. Usually made from the decay products of animal waste.

Nox : Night.

Nux moschata : Nutmeg.

Nycthemeron : See dies et nox.

Obulus scrupulus semis : An obulus is half a scruple.

Oleum. — — — — Oel.
Oleum commune,
oleum oliuarum,
gremiale. — Baum=Oel.

Oleum Saturni. — — — Bley=Oel.

- - Sulphuris. Schwefel=Oel.

- - Talchi oder
Talci. — — —Talck=Oel.

- - Tartari Sen-
nerti. D. Sennerts Wein-
stein=Oel.

- - Vitrioli. — — —Vitriol=Oel.

Ouum. — —Ein Ey.

Pars cum parte. Eine Maſſe von
gleichviel Gold
und Silber unter-
einander cämen-
tirt und graduirt.

Per deliquium. — — — —Von ſelbſt zerfloſſen.

Phlegma, aqua inſi-
pida. Ein unſchmackhaftes
Waſſer.

Piſcis, ichthys. — — —Ein Fiſch.
Plumbago, plum-
bum album, ſiehe
Ceruſſa. Bleyweis.

Plumbum, Satur-
nus. Bley.

46

Oleum : Oil.

Oleum commune, oleum olivarum, gremiale : Common oil. Olive oil.

Oleum Saturni : A solution of lead chloride $PbCl_2$ in spirit of salt HCl, sometimes also called butter of Saturn.

Oleum Sulphuris : An oil made from sulphur.

Oleum Talchi or Talci : Made by calcining talc. Talc powder readily absorbs oils.

Oleum Tartari Sennerti : Watery solution of Potassium carbonate or tartrate. First mentioned by Daniel Sennert.

Oleum Vitrioli : Oil of vitriol, concentrated sulphuric acid H_2SO_4.

Ovum : Egg.

Pars cum parte : Part with part, take an equal part of each substance.

Per deliquium : Dissolved into liquid through absorbing moisture from the air.

Phlegma, aqua insipida : Phlegm, one of the four humours. A tasteless water.

Piscis, ichthys : Fish.

Plumbago, plumbum album : See Cerussa.

Plumbum, Saturnus : Lead.

Praecipitatio, prae-cipitatus.	≡ – – – – –	Niedergeschlagen, gefällt.
Praeparatio, prae-paratus.	≠≠ppt – – – – –	Präparirt.
Pugillus.	P. ꝟ꞊ꞃꝓ p.– –	Ein Pugill, was man zwischen 3. Finger fassen kan.
Pugillus semis.	Pβ ꝟ꞊ꞃꝓ/p./β. –	Eine halbe Pugill, oder auch so viel man zwischen 2. Finger fassen kan.
Puluis.	Pulv. ⚵ ⚶ ⚴ Ҳ	Ein Pulver.
Puluerisare.	A̷ ⚴ ⚵ H̷	Zu Pulver zerstossen.
Purificatio.	♅ ♅ ♅ ♅ ♅	Die Reinigung.
Putredo, putrefa-ctio.	Ψ ✝ Ψ E.S. ▷	Die Fäulung, das Verfaulen.

Q.

Quantum placet.	q. pl. – – – –	So viel beliebt.
Quantum satis.	q.s. – – – –	Bis es genug ist.
– – – uis.	q.v. – – – –	So viel man will.
Quinta essentia.	2. E. – – –	Die Quint-Essenz, das feineste und beste.

48

Praecipitatio, praecipitatus : A precipitate cast down in the reaction. To precipitate.

Praeparatio, praeparatus : A preparation. To prepare.

Pugillus : A pinch, a quantity that can grasped in the fingers.

Pugillus semis : Half a pinch.

Pulvis : Dust or powder.

Pulverisare : To pulverise, or reduce to a powder.

Purificatio : To purify.

Putredo, putrefactio : Rotten or putrid. To make rotten.

Quantum placet : As much as you please.

Quantum satis : As much as is sufficient. Just enough.

Quantum vis : As much as you wish.

Quinta essentia : Quintessence.

R.

Radix, radices. *Rad.* – – –Wurzeln.

Rasura, raspatum. *Ras. R. rasur. rasp.* Etwas Geraspeltes.

Realgar, fumus, exhalatio & concretio. Ein Rauch, der sich wieder in eine trockene Materie zusammengesetzt hat.

Receptaculum, Recipiens, excipulum. Ein Recipient oder Glas, das man bey Destillationen vorschlägt, um das herübergehende aufzufassen.

Receptum, formula medica, recepta. *Recept.* Ein Recept.

Recipe. Nimm, NB. wird vornenhin auf den Recepten gesetzt.

Reductio. Die Reduction, oder Wiederbringung in die vorige Gestalt.

Regulus. – –Ein Metall-König.

Regulus antimonii medicinalis. – – – – –Der Arzney Spiesglas-König.

Renouatio metallorum. Die Erneurung der zerstörten Metalle.

50

Radix, radices : Root or roots.

Rasura, raspatum : To scrape, rasp, grate or shave off parts.

Realgar, fumus, exhalatio et concretio : Red Arsenic sulphide As_4S_4.

Receptaculum, Recipiens, excipulum : Receptacle, receiver, vessel for holding liquids.

Receptum, formula medica, recepta : A prescription, medical formula, or receipt.

Recipe : Recipe or formula for making something.

Reductio : Reduction.

Regulus : An ore reduced to its pure metallic form.

Regulus antimonii medicinalis : Metallic antimony.

Renovatio metallorum : The renewing or re-extraction of corroded or oxidised metals.

Refina.	⊡ _ _ _ _	—Ein Harz.
Retorta, cornuta, matracium.	6, 6, 6, 6 _ _	Eine Retorte, Elephanten = Schnabel.
Reuerberatio.	ℛ ℓ𝒻 ⇄ Z₂ R	Das Reverberiren.
Reuerberatorium, reuerberium.	𝄐, ⊥ _ _ _	Ein Reverberir = Ofen.
Rhabarbarum.	*Rhab.* _ _ _	Rhabarbara.
Rifigallum, fiehe Auripigmentum.		

S.

Saccharum.	Σ _ _ _ →	Zucker.
Sal.	θ _ _ _ _	Salz.
Sal alcali, oder alkali, f. alcali und Cineres clauellati.	⊤, ⚥⊦, ⊡, ♄ X, 8, G, ⊞, F, E 𝄐, ⊡, ⊓̄, 𝘈, 𝘈 R, ⚛, ⚥, ⊡, ⊖4 L A ⊠, ⊓̇, ⊙̇, ℛ,	Laugen=Salz, Potafche.
Sal ammoniacum oder armoniacum.	⊕ℰc. ✳ ⊙ K Ψ ✳ F, ∞, Z, ✕, S ⊃C, ⊃IC, ⋯ ✳ X, ♄ X, ⚶, X M, ✳, ⊥, ◇, ⊥ X, ⊃C,	Salmiack.

Resina : Resin.

Retorta, cornuta, matricium : Retort, flask with a beak, matrass.

Reverberatio : Expose a substance to naked flame.

Reverberatorium, reverberium : A furnace of reverberation, in which a substance is in direct contact with the burning coals or wood.

Rhabarbarum : Rhubarb.

Risigallum : See Auripigmentum.

Saccharum : Sugar.

Sal : Salt.

Sal alcali or alkali : See Alcali and Cineres clavellati.

Sal ammoniacum or armoniacum : Sal ammoniac or ammonium chloride NH_4Cl.

Sal commune.   Gemein Saltz, Kus
                                           chen = Saltz.

Sal colcotharium,   — — — —Vitriol=Saltz.
   vitriolum vomi-
   tiuum.

Sal essentiale vini,
   siehe Terra folia-
   ta tartari.

Sal gemmae, oder   Stein=Saltz.
   fossile, indum.

Sal marinum.   — — —Meer = Saltz.

Sal petrae, aphro-
   nitrum, flos pa-   Mauren = oder Kel-
   rietis, faex nitri,   ler = Salpeter.
   Nitrum Graeco-
   rum, Nitrum sto-
   lidum.

54

Sal commune : Common salt, Sodium chloride NaCl.

Sal colcotharium vitriolum vomitiuum : The brownish red peroxide of iron Fe_2O_3 which remains in the retort after the distillation of sulphuric acid from iron sulphate.

Sal Essentiale vini : See Terra foliata tartari.

Sal gemmae or fossile : Rock salt.

Sal marinum : Sea salt.

Sal petrae, aphronitrum, flos parietis, faex nitri, Nitrum Graecorum, Nitrum stolidum : Saltpetre, Potassium nitrate KNO_3.

Sal Tartari fixum.	⚗ ⚗ ⚗ ⚗ ⚗ ⚗ ⚗	Weinstein=Salz.
Sal essentiale,f.Ter- ra foliata tartari. Sal vini essentiale,	⚗ ⚗ ⚗	ist wieder das vorige.
Sal volatile.	θV. θΛ. — —	Ein flüchtig Salz.
Sal urinae.	⊞, ⊕, ▭, ⊟	Urin, Harn=Salz.
Sapo.	◇ ◇ — — —	—Seiffe.
Saturnus , f. plum- bum.	⊡, ⟟, 5, 5, ♄ ♓, ♋,	Bley.
Scriptulus undScri- pulus,		ist so viel als Scru- pulus.
Scrupulus.	Ɉ, Ɔ, Ϝ — — —	Ein Scrupel , 20. Gran.
Secundum artem.	f. a, f. A —	— Nach der Kunst.
Semen, Semina.	Sem. — — —	—Saamen.
Semis, femissis.	S. ß. — — —	Halb.
Semiuncia, femun- cia , uncia femis oder dimidia.	Ʒ ß. Ʒiv — —	Eine halbe Onz, ein Loth oder 4. Quintlein.
Sextarius.	♄, Ew, ♄, — —	Ein Sextarius.

Sal Tartari fixum : Tartar, Potassium hydrogen Tartrate $KC_4H_5O_6$, usually made from the precipitate thrown down from wine.

Sal essentiale : See Terra foliate tartari, sal vini essentiale.

Sal volatile : Ammonium carbonate $(NH_4)_2CO_3$.

Sal urinae : Salt made from urine, probably ammonium chloride.

Sapo : Soap.

Saturnus : Saturn, see Plumbum.

Scriptulus and Scripulus: Alternative names for Scrupulus.

Scrupulus : A weight equivalent to 20 grains, or a twentieth of an ounce.

Secundum artem : Following the art.

Semen, semina : Seed.

Semis, semissis : Half.

Semiuncia, semuncia, uncia semis or dimidia : Half an ounce.

Sextarius : Ancient Roman measure of about a pint (500 ml). This was a sixth of a congius (about seven pints).

Siccare.	𝒟. — — — —	Trocknen.
Siccum.	ſ — — — —	Trocken.
Signa, ſignetur.	ſ — — — —	Ueberſchreibe es.
Simplex & compoſitum.	ſ. et C. — —	Einfach und zuſammengeſetzt.
Sine vino.	ſ.v. — — —	Ohne Wein.
Sine ſtipitibus.	ſ.ſ. — — — —	Ohne Stiele.
Soda.	⊔ — — —	Spaniſcher Sod.
Sol, ſ. aurum.		
Solutio, ſoluere.		Das Solviren, Auflöſen.
Species.	ſpec. — — —	Species.
Spiritus.	ſp. ſpir. — —	Ein Geiſt.
Spiritus vini.		Brandwein.
Spiritus vini rectificatiſſimus, ſiehe Alcohol vini.		
Stannum, ſiehe Jupiter.		

58

Siccare : To dry.

Siccum : Dry. One of the Four Qualities.

Signa, signetur : A sign or symbol. It is indicated by a sign.

Simplex et compositum : Simple uncompounded substance and a compounded mixture.

Sine vino : Without wine.

Sine stipitibus : Without stems. Thus only the leaves or flowers of a plant.

Soda : Sodium carbonate Na_2CO_3.

Sol : See aurum.

Solutio, solvere : A solution. To dissolve into a solution.

Species : Look or examine. Appearing the same.

Spiritus : Spirit.

Spiritus vini : Spirit of wine Ethyl alcohol C_2H_5OH.

Spiritus vini rectificatissimus : Rectified, distilled or concentrated Ethyl alcohol.

Stannum : See Jupiter.

Stratum super stratum. $SSS.$ ⚗, $SSS‡$ Ein Stratum super Stratum.

Sublimatio, sublimare. ⚗ Sublimiren, sublimirt.

Succinum album, Leucelectrum. $BS,$ $SVA,$ $S.V.A$ —Weiſer Agtſtein.

Succus. ♃ — — — —Ein Saft.

Sulphur. 🜍 🜍 🜍 🜍 🜍 Schwefel.
🜍 🜍 🜍 🜍 🜍 🜍
🜍 🜍 🜍 🜍 🜍
🜍 🜍 🜍

Sulphur nigrum. 🜍 🜍 🜍 Der ſchwarze Schwefel.

Sulphur philoſophorum. 🜍 🜍 🜍 🜍 — Der philoſophiſche Schwefel.

Sulphurſtillatitium. 🜍 — — — —Tropf-Schwefel.

Sulphur tartari, tinctura ſulphuris. 🜍 — — — —Weinſtein-Tinctur.

Sulphur viuum. 🜍 🜍 — — Lebendiger Schwefel.

60

Stratum super stratum : Layer upon layer.

Sublimatio, sublimare : Sublimate. To sublime.

Succinum album, Leucelectrum : White amber.

Succus : Juice.

Sulphur : Sulphur.

Sulphur nigrum : Black sulphur. The impure crude greyish residue left from the purification or sublimation of sulphur from ores.

Sulphur philosophorum : The sulphur of the philosophers.

Sulphur stillatitium: Distilled sulphur.

Sulphur tartari, tinctura sulphuris : The sulphur of tartar or an oil made from tartar.

Sulphur vivum : Living sulphur, or that found in Nature, say around volcanic vents.

T.

Talca, Talcum.	[alchemical symbols]	Talch, Talck.
Tartarus.	[alchemical symbols]	Weinstein.
Tartarus emeticus.	[alchemical symbols] — — — —	—Brech-Weinstein.
Tauri priapus.	[alchemical symbol] — — — —	Ein Ochsenzimmer, oder Farrenschwantz.
Terebinthina.	[alchemical symbols] — — — —	—Terpenthin, Cläret, Lerchen-Hartz.
Terra.	[alchemical symbols]	Erde.
Terra foliata tartari.	[alchemical symbols]	Weinstein-Saltz mit Essig geträncket.
Terra Lemnia.	[alchemical symbols]	Gesiegelte Erde von der Insel Lemno.

62

Talca, Talcum : Talc, a friable mineral, which can be easily powdered.

Tartarus : Tartar, the stony material deposited from wine. Consists primarily of Potassium hydrogen tartrate, $KC_4H_5O_6$.

Tartarus emeticus : Potassium antimonyl tartrate $K_2Sb_2(C_4H_2O_6)_2$. A powerful emetic. Sometimes made by leaving wine in a cup cast of pure antimony for a day or so.

Tauri priapus : Powder of bull's penis.

Terebinthina : Turpentine, a volatile oil distilled from the Terebinth tree.

Terra : Earth.

Terra foliata tartari : Potassium acetate CH_3COOK prepared from potassium carbonate and vinegar.

Terra Lemnia : A red (or sometimes white) clay from the Isle of Lemnos. The clay contains a high percentage of iron oxides which give it a reddish brown colour. It was used in medication primarily for skin diseases.

Terra sigillata alba. ⌖ ⌖ ⑤ — — Weiſſe geſiegelte Er-
de.

Tigillum, ſ. Cruci-
bulum. ⌖ Ein Tiegel.

Tinctura. ⌖ Eine Tinctur.

Turbithum, Turpe-
thum minerale. ⌖ — — — —Mineral-Turpeth.

Tutia Alexandrina. ⌖ Tuviii, Tutſus,
Tutia Officinarum, grauer Hütten-
Cadmia factitia, rauch.
Cadmia forna-
cum.

U.

Uncia. ℥, 8ß, 33, ℈ Eine Ontz, 2. Loth,
8.Quentlein.

Uncia ſemis. ℥ß — — —Eine halbe Ontz, ein
Loth.

Urina, lotium. □, ⌖ Urin, Harn.
□, ⌖

Terra sigillata alba : White clay.

Tigillum : See Crucibulum.

Tinctura : Ticture.

Turbithum, Turpethum minerale : Turpeth can be the root of the turpeth plant used as an emetic. It can also be Turpeth mineral, a lemon yellow powder, basic mercuric sulphate $Hg_3O_2(SO_4)$ made by boiling mercury and sulphuric acid together to dryness, and throwing the resulting mass into boiling water.

Tutia Alexandrina : A clayey ore of zinc found in Persia.

Tutia Officinarum, Cadmia factitia, Cadmia fornacum : Zinc. Usually extracted from the soot found in flues of lead or antimony furnaces.

Uncia : Ounce.

Uncia semis : Half an ounce.

Urina, lotium : Urine.

V.

Vaporis balneum, siehe Balneum roris.

Venus cuprum, siehe auch aes.   Kupfer.

Ver.   — — —Der Frühling.

Vesica destillatoria.   — — —Eine Destillir-Blase.

Vinum.   — —Wein.

Vinum adustum, spiritus frumenti.   — — —Frucht-Brandwein.

Vinum album.   Weisser Wein.

Vinum alcalisatum, oder circulatum, correctum,   ist so viel als Alcohol vini.

Vinum emeticum.   — — — — —Ein Brech-Wein.

Vinum Hippocraticum.   — — Ein Hippocras-Wein.

Vinum medicatum.   — — — —Ein Kräuter-Wein.

Vinum mortuum, s. acetum.   — — —Essig.

Vinum rubrum.   — — — — Rother Wein.

Vaporis balneum : See Balneum roris.

Venus cuprum : See aes.

Ver : Spring.

Vesica destillatoria : Distillation vessel.

Vinum : Wine.

Vinum adustum, spiritus frumenti : Distilled wine, brandy.

Vinum album : White wine.

Vinum alcalifatum or circulatum, correctum : Another name for alcohol made from wine.

Vinum emeticum : Antimonated wine. Wine acted upon by mercury to create Tartar emetic.

Vinum Hippocraticum : Hippocras. Wine flavoured with spices.

Vinum medicatum : Medicinal wine.

Vinum mortuum : See acetum.

Vinum Rubrum : Red wine.

Viride aeris, viride graecum, oder hispanicum. Grünspan.

Vitellus, Luteum, Luteum oui. Eyerdotter, das Gelbe des Eys.

Vitriolum, f. auch atramentum. Vitriol, Kupfer-Waſſer.

Vitriolum album. Weiſſer Vitriol, Caliṫel-Stein.

Vitriolum Romanum. Römiſcher Vitriol.

Vitrum. Glas.

Vitrum antimonii, ſiehe antimonii vitrum.

Volatile. Flüchtig.

Zincum, Zinctum, Zinck, Zink, ſiehe Marcaſita aurea.

Zingiber.
Zinziber. Ingber.

Viride aeris, viride graecum or hispanicum : Verdigris.

Vitellus, Luteum, Luteum ovi : Lute made from egg yokes.

Vitriolum : See atramentum.

Vitriolum album : White vitriol. Zinc sulphate $ZnSO_4$.

Vitriolum Romanum : Roman vitriol. Copper sulphate $CuSO_4$.

Vitrum : Glass.

Vitrum antimonii : Glass of antimony. A vitrified mixture of oxides and sulphides of Antimony.

Volatile : Volatile.

Zincum, Zinctum, Zinck, Zink : See Marcasita aurea.

Zingiber, Zinziber : Ginger.